LIFE SCIENCE IN DEPTH

INHERITANCE
AND
SELECTION

Ann Fullick

Heinemann
LIBRARY

www.heinemann.co.uk/library
Visit our website to find out more information about Heinemann Library books.

To order:

 Phone 44 (0) 1865 888066

Send a fax to 44 (0) 1865 314091

Visit the Heinemann bookshop at www.heinemann.co.uk/library to browse
our catalogue and order online.

First published in Great Britain by
Heinemann Library, Halley Court, Jordan Hill,
Oxford OX2 8EJ, part of Harcourt Education.

Heinemann is a registered trademark of
Harcourt Education Ltd.

© 2006 Harcourt Education Ltd.

Editorial: Sarah Shannon and Dave Harris
Design: Richard Parker and Q2A Solutions
Illustrations: Q2A Solutions
Picture Research: Natalie Gray
Production: Chloe Bloom

Originated by Modern Age Repro
Printed and bound in China by South China
Printing Company

10 digit ISBN: 0 431 10899 4
13 digit ISBN: 978 0 431 10899 5

10 09 08 07 06
10 9 8 7 6 5 4 3 2 1

British Library Cataloguing in Publication Data
Fullick, Ann, 1956-
  Inheritance and selection.
  - (Life science in depth)
  576.5
A full catalogue record for this book is available
from the British Library.

Acknowledgements
The publishers would like to thank the following
for permission to reproduce photographs:
Alamy pp. **47** (Cosmo Condina), **45** (Grace
Davies), **58** (PhotoLibrary Wales), **55** (Robert
Harding World Imagery); Corbis pp. **30**
(Bettmann), **9** (Joe McDonald), **34** (Kennan
Ward), **40** (Lester Lefkowitz), **28** (Melissa
Moseley/Sony Pictures/Bureau L.A. Collections),
49 (Najlah Feanny), **57** (Sygma), **4**; Empics
p. **27** (Rebecca Naden/PA); Frank Lane Picture
Agency pp. **39** (Foto Natura Stock), **44** (Terry
Whittaker); Getty Images pp. **5** (Philip Lee
Harvey), **32** (Robert Hardding World Imagery),
17 (Stephen Johnson); John Cole p. **43**; Michael
Greenlar p. **24** (Topham/Image Works); Science
Photo Library pp. **53** (BSIP, Laurent), **8a**, **8b**, **29**
(CNRI), **23** (Darwin Dale), **1** , **7**, **36** (Dr Gopal
Murti), **48** (Ed Young/Agstock), **20** (James King-
Holmes), **10** (Philippe Plailly/Eurelios), **33**.

Cover photograph of a DNA molecule, reproduced
with permission of Science Photo Library.

Our thanks to Emma Leatherbarrow for her
assistance in the preparation of this book.

Contents

Words printed in the text in bold, **like this**, are explained in the Glossary.

A family likeness?

Cats have kittens, people have babies, and oak trees produce acorns that grow into little oak trees. Young animals and plants usually look similar to their parents, and to other animals or plants of the same type. In fact, many of the smallest **organisms** that live in the world around us are actually identical to their parents.

PASSING IT ON

Family likenesses are often a topic of conversation. People might say "you've got your grandfather's good looks" or "I see you've got the family ears". Characteristics like these are **inherited**, passed on from parents to their children through **reproduction**. Most families have characteristics that can be clearly seen from generation to generation – and the same thing is true of other animals and plants.

ONE PARENT OR TWO?

Reproduction is a key process in life. It is during reproduction that **genetic information** is passed on from parents to their offspring. There are two very different ways of reproducing, which are called **asexual reproduction** and **sexual reproduction**.

A mass of daffodils like this can contain hundreds of identical flowers grown from bulbs that reproduce asexually.

Asexual reproduction involves only one parent, which produces new organisms completely identical to itself. There is no variety, but it is very easy. Asexual reproduction does not require a partner with whom to share genetic information. It is very common in the smallest animals and plants, and in **bacteria**, but many bigger plants like daffodils, strawberries, and brambles reproduce asexually, too. Asexual reproduction also takes place all the time in your own body, such as when cells divide to grow and to replace worn out tissues.

The other way of passing information from parents to their offspring is through sexual reproduction. Each parent has a special sex cell called a **gamete**, which joins together with the other parent's sex cell to form a new individual. A living thing that is formed by sexual reproduction will inherit genetic information from both parents. It will inherit some characteristics from each parent, but will not be exactly like either of them. In animals the special sex cells involved in sexual reproduction are known as **ova** (eggs) and **sperm**. In plants they are called **ovules** and **pollen**.

THE MESSAGE IN THE CELLS

Information passed from generation to generation during reproduction is carried by units of inheritance called **genes**. They are made up of a special chemical known as **DNA** (deoxyribonucleic acid). Unravelling the secrets of DNA and our growing understanding of how inheritance works have led to some of the most exciting developments of 20th and 21st century science.

Any family group is made up of individuals who all look different. Yet because of their shared DNA there are many similarities between them. Some of these similarities can be seen in their appearance while others are hidden away in the working of their cells.

Inheritance

When sexual reproduction takes place, information from two parents is mixed to make a completely new offspring. Half of the information comes from the male sex cell (gamete) and half comes from the female gamete. Each gamete formed is unique, with a slightly different combination of genes from any other. This means that from hundreds or even thousands of offspring, no two will be exactly the same. The only exception is identical twins, which are formed when a fertilised egg splits to make two genetically identical individuals.

THE BLUEPRINT OF LIFE

Almost all of the cells of your body – with the exception of your **red blood cells** – contain a **nucleus**, which is the "control room" of the cell. It contains all the plans for making a new cell – and what is more, the blueprint for a whole new you!

Imagine the plans for building a car. They would cover many sheets of paper. Yet in every living organism, the nucleus of the cells contains the information required to build a whole new organism. A human being is far more complicated than a car, so how does all this information fit?

Inside the nucleus of every cell there are thread-like structures called **chromosomes**. This is where the genetic information passed on from parent to child is stored. The chromosomes are made up of DNA, and this amazing chemical (see page 10) carries the instructions to make all the **proteins** in your cells. Many of these proteins are actually **enzymes**. Enzymes control the production of all the chemicals that make up your body. They also affect what you look like and who you are.

CLASSIC EXPERIMENT seeing the chromosomes

Most of the time you cannot see the chromosomes in the nucleus of a cell. However, cells divide regularly in order for the organism to grow and also to replace worn out cells. When a cell is dividing in two, the chromosomes get much shorter and thicker. At this stage they will take up special colours called stains which make them much easier to see under the microscope. In fact the name "chromosome" means "coloured body" – and it refers to what the chromosomes look like when they have taken up the stain. This experiment was done for the first time by Walther Fleming, a German scientist, and now is carried out regularly in school laboratories across the world.

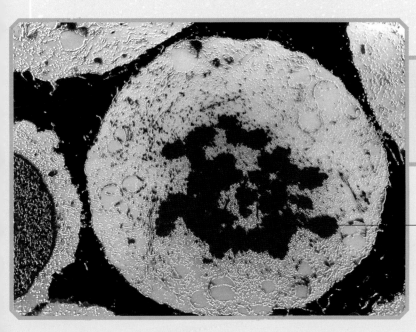

These cells are in the growing root tip of a plant, so they are dividing rapidly. The chromosomes have taken up the red stain and so you can see them clearly.

chromosomes

Did you know..?

A single female herring can lay between 20,000 and 200,000 eggs at a time – and genetically, every one of them will be slightly different!

MORE ABOUT CHROMOSOMES

Each different type of organism has a different number of chromosomes in its body cells. For example, humans have 46 chromosomes and tomatoes have 24, while turkeys have 82! You inherit half your chromosomes from your mother and half from your father, so chromosomes come in pairs. Humans have 23 pairs, tomatoes 12 pairs, and turkeys 41 pairs of chromosomes.

Scientists can photograph the chromosomes in human cells when they are dividing. They can then arrange them in pairs to make a special picture known as a **karyotype**.

BOY OR GIRL?

Human karyotypes show 23 pairs of chromosomes. In 22 of the pairs, the chromosomes are the same size and shape, whether you are a male or a female. These 22 pairs of chromosomes are known as the **autosomes**. They control almost everything about the way you look and the way your body works.

The remaining pair of chromosomes is different for males and females. These are known as the **sex chromosomes** because they decide whether you are male or female. Everyone inherits an **X chromosome** from their mother's egg. If this joins with a sperm carrying another X chromosome, the baby will be a girl. If the egg is fertilized by a sperm carrying a **Y chromosome**,

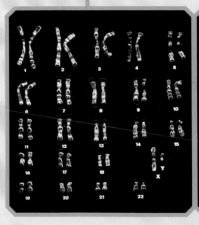 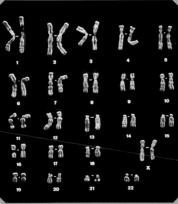

Karyotypes like these of a man (on the left) and woman (on the right) have helped scientists find out more about the mysteries of inheritance.

the baby will be a boy. X chromosomes provide information about female characteristics, but they also carry information about lots of other things such as the way your blood clots and the formation of your teeth, body hair, and sweat glands. Y chromosomes carry information about male characteristics and just a few other bits of information.

RINGING THE CHANGES

In mammals, including humans, females are XX and males are XY, but this is not the case in all animals. In birds, for example, females are XY and the males are XX. For reptiles and amphibians, things are even more complicated. In some **species** the males are XY and the females XX, while for others it is the opposite.

RECENT DEVELOPMENTS

In the last fifteen years, research has shown that some types of animals, such as crocodiles and tortoises, do not have any sex chromosomes. Instead, their sex is decided by the temperature of the eggs while they develop. In most tortoises, males develop at cooler temperatures and females at warmer temperatures. Alligators produce males at warm temperatures and females at cool temperatures. The eggs of crocodiles and snapping turtles develop into females at both cold and hot temperatures, and into males at temperatures in between.

A few degrees change in temperature makes all the difference for the eggs of the spur-thighed tortoise.

THE MIRACLE MOLECULE

The chromosomes you inherit from your parents carry all the information needed to make a new you in the form of genes. Each gene is a small section of DNA. It seems amazing to think we have only understood this for about 50 years.

DNA is a long **molecule** made up of two strands twisted together to make a spiral. This is known as a double helix, which looks like a ladder that has been twisted round. The big DNA molecule is actually made up of lots of smaller molecules joined together. These include four different molecules called **bases**, which appear time after time in different orders but always paired up in the same way. These bases link the two strands of the DNA molecule together. Genes are made up of repeating patterns of bases.

The structure of the DNA molecule takes you deep into the chemistry of life. A small change in the arrangement of bases in your DNA would have meant a very different you!

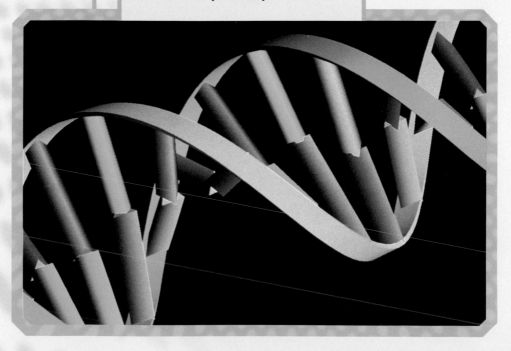

SCIENCE PIONEERS cracking the code

By the 1940s, most scientists had decided that DNA was probably the molecule that carried inherited information from one generation to the next. But how did it work? To understand inheritance, scientists needed to understand the structure of DNA, but no-one had managed to work that out.

By the 1950s, two teams in the UK were getting close. Maurice Wilkins and Rosalind Franklin in London were taking special X-ray photographs of DNA. They were looking at the patterns in the X-rays in the hope that these patterns would show them the structure of the molecule. It was very tricky and the pictures, when they got them, were very difficult to understand.

At the same time, James Watson (a young American) and Francis Crick (from the UK) were also working on the DNA problem at Cambridge. They took all the information they could find on DNA – including the X-ray pictures from London – and tried to build a model of the molecule that would explain everything they knew. When they finally realised that the bases always paired up in the same way they had cracked the code. The now famous double helix was seen for the first time.

Since the structure of DNA was revealed, there has been an enormous increase in the amount of research done in the field of genetics.

Did you know..?

The bases which make up DNA are called adenine, thymine, guanine, and cytosine. Adenine always pairs up with thymine, and guanine pairs with cytosine. These pairs are vital to the shape of the DNA molecule and the way it works.

ALL IN THE GENES

The genes found on the chromosomes control everything that happens in your cells. They do this by organizing the **amino acids**, which make up all the proteins you need. A gene can be tens of thousands of bases long, but the bases are always arranged in groups of three. Each group of three bases provides the code for one amino acid. The order of the bases in the DNA acts as another code. This code tells the cell in what order to join up the amino acids to make a particular protein. A protein is a long chain of amino acids. Different combinations of amino acids can be joined together to make different proteins.

The genes in the nucleus of one of your cells control the proteins that are made, which in turn affect almost everything about you.

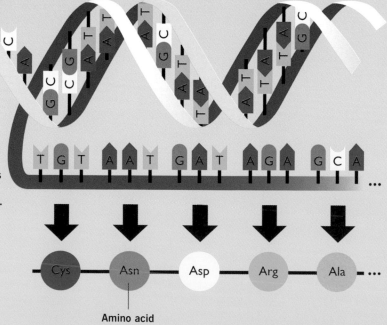

1 DNA unzips.

2 DNA sequence forms the genetic code, where each group of 3 letters is the code for a specific amino acid.

T G T   A A T   G A T   A G A   G C A   ...

3 Growing protein chain made up of amino acids linked together.

Cys — Asn — Asp — Arg — Ala — ...

Amino acid

HOW PROTEINS ARE MADE

DNA is found in the nucleus of the cell, but proteins are made outside the nucleus in the **cytoplasm**. There are holes in the membrane of the nucleus but DNA is a very big molecule and it cannot get through them – so how does it control the protein-making process?

When a new protein is needed, a small bit of DNA unzips to reveal the right gene. A copy of this gene is made using a chemical called RNA (ribonucleic acid). RNA is very similar to DNA except it contains the base uracil instead of thymine. The copy can be made because the bases still pair up in the same way, with uracil pairing up with adenine. The RNA copy of the single gene is small enough to pass out of the nucleus into the cytoplasm of the cell, where it is used to build up a chain of amino acids. These eventually form the protein you need. This amazing process is known as protein synthesis.

A lot of the original work breaking the code between the RNA bases, DNA, and amino acids was done on bacteria. Research that has been done since shows that the code is very similar for plants and animals, including humans.

THE GENETIC DICTIONARY

Once scientists understood how the information in your genes causes the freckles on your skin or your curly hair, they set out to try and understand which combinations of DNA bases coded for which amino acids – in other words, the genetic alphabet. There are only about 20 amino acids, but it was still a very difficult job. Cracking the amino acid code was an important step towards another, bigger goal – working out the full code of human DNA.

Did you know..?

By the end of 2004, the exact number of genes in the **human genome** was still not known for sure, but best estimates put it between 20,000 and 25,000.

THE HUMAN GENOME PROJECT

The more scientists learned about DNA, genes, and the way characteristics are inherited, the more important it became to try and unravel the whole of the human DNA – known as the **human genome**.

Scientists believed that understanding the human genome was the key to the mystery of human life itself. The Human Genome Project was set up in 1989 firstly to identify all of the genes in the human chromosomes, and then to discover

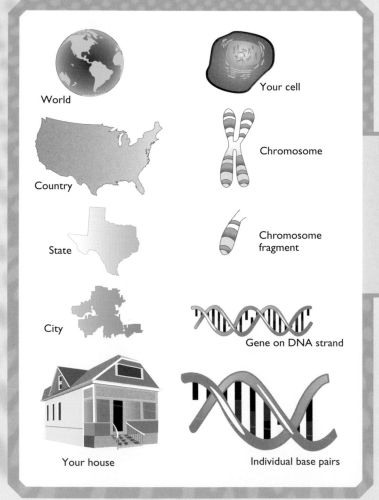

World

Your cell

Country

Chromosome

State

Chromosome fragment

City

Gene on DNA strand

Your house

Individual base pairs

Working out the bases of the human genome is a bit like taking a map of the world and pinpointing your personal address. Eventually it will make it possible for you to know your own genetic makeup.

the 3 thousand million (3,000,000,000) base pairs which make up the human DNA. Scientists in eighteen different countries worked on different bits of the genome at the same time so they could announce the results in the year 2000. The Human Genome Project has cost around US $2.7 billion, and has shown that everyone shares at least 99.99 percent of their DNA with everyone else.

When the Human Genome Project was first launched, scientists wanted to find the position of all the individual genes on the human chromosomes. Then, as technology developed, they moved on to looking at the individual bases themselves. They thought it would take 15 years to complete the project, but technology improved so quickly that the genome was ready two years ahead of schedule. Now we know all of the bases, all that is left is to work out exactly what all the different genes do.

SCIENCE PIONEERS making the Genome Project possible

The Human Genome Project worked so well because the scientists had the right tools for the job. This was largely due to scientific pioneers, called Frederick Sanger and Kary Mullis.

Frederick Sanger, a British **biochemist**, is one of a small number of scientists with two Nobel prizes, the second of which was for developing a way of working out the sequence of bases in DNA. It used to be done by hand, base by base. But now machines can sort out hundreds of DNA bases at a time!

Kary Mullis is an American scientist who, like Sanger, won a Nobel prize. He developed a way of making millions of copies of a strand of DNA quickly in the lab, called the Polymerase Chain Reaction. This was really important for the Human Genome Project because it gave scientists plenty of identical DNA to work on.

Passing it on

Do you have much in common with the rest of your family? Sometimes there are strong family likenesses, and sometimes everyone looks very different. Whatever you are like, the key is in your genes.

THE GENE LOTTERY

Each of your chromosomes carries thousands of genes, arranged so that both chromosomes in a pair carry genes controlling the same things in the same positions. So not only do your chromosomes come in pairs but also your genes – one from each parent.

Each pair of genes controls different things about you but the genes in a pair can come in different forms. These different versions of the same gene are called **alleles**. Some of our characteristics are controlled by one gene with just two possible alleles. We can use these genes to help us understand how inheritance works. There are genes that decide whether your earlobes are attached closely to the side of your head or hang freely, or whether you have dimples when you smile. The gene that controls dimples has two possible forms – an allele for dimples, and an allele for no dimples.

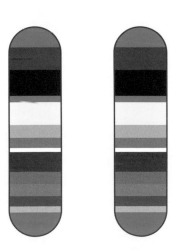

The pairs of genes on your chromosomes interact to control most things about your body. The genes are shown as coloured bands on this chromosome diagram.

The gene for dangly earlobes also has two possible alleles – one for dangly earlobes and one for attached earlobes. You will get a mixture of alleles from your parents, which is why you don't look exactly like either of them.

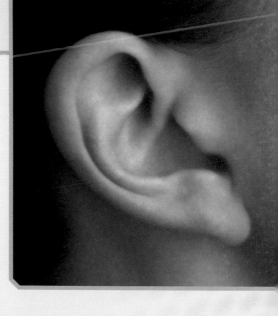

This person has inherited the gene for dangly earlobes. Physical characteristics like this depend on which alleles are passed on from the parents.

HOW DOES IT WORK?

Imagine a bag containing marbles. If you put your hand in and, without looking, picked out a pair of marbles, what might you get? If the bag contained only red marbles or only blue marbles, the pairs would all be the same. But if the bag held a mixture of red and blue marbles you could end up with three possible pairs – two blue marbles, two red marbles, or one of each. This is what happens when you inherit genes from your parents. For example, if both of your parents have two alleles for dimples, you will definitely inherit two dimple alleles and you will have dimples. If both of your parents have two alleles for no dimples, you will inherit alleles for no dimples and you will be dimple free. But if your parents both have one allele for dimples and one for no dimples, you could end up with two dimple alleles, two no-dimple alleles, or one of each!

Did you know..?

There is one thing that definitely is not controlled completely by your genes – your fingerprints. Even identical twins have different fingerprint patterns, showing that something else must also be involved in making your fingerprints.

DOMINANT OR RECESSIVE?

If you inherit an allele for dimples from your father and an allele for no dimples from your mother, how does your body know whether to produce dimples or not? In most cases one allele will dominate the other. In the case of dimples, the allele for dimples will always triumph over the allele for no dimples, so if you get one of each, you will have dimples.

Alleles like this are known as **dominant** alleles. This means that you will have that characteristic even if you inherit the allele from only one parent. We represent a dominant allele using a capital letter, so the allele for dimples is "D" while no dimples is "d". You only have to inherit one dimple allele to have dimples. Think back to the marbles – if the red marbles represent the "D" allele, you will have dimples whether you pick two reds or a red and a blue. The allele that is not dominant is called **recessive**. You will only show the characteristic if you inherit a recessive allele from both parents – in this case, picking a pair of blue marbles.

If you inherit two identical alleles – either two dominant or two recessive alleles – you are said to be **homozygous** for that gene (both the alleles are the same). If you inherit two different alleles (one recessive and one dominant), you are **heterozygous** for that gene. Your arrangement of alleles is known as your **genotype**. The way you look as a result of your genes is called your **phenotype**.

DRAWING IT OUT

If a parent has only one form of an allele (two recessive alleles or two dominant alleles), he or she can only pass on that form of the gene. But if he or she has two different alleles, half of the eggs or sperm will contain the dominant allele and half will contain the recessive allele. This has a big effect on what the child might be like.

The way different alleles are passed on from one generation to another can be illustrated in a **genetic diagram**. These diagrams can be used to work out what the offspring of any parents might be like.

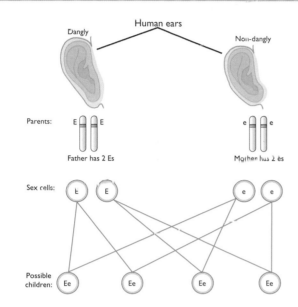

Human ears

Dangly   Non-dangly

Parents: E ⬛⬛ E   e ⬛⬛ e

            Father has 2 Es   Mother has 2 es

Sex cells: E   E   e   e

Possible
children: Ee   Ee   Ee   Ee

All the children will have dangly ear lobes, but will carry a recessive allele for attached lobes.

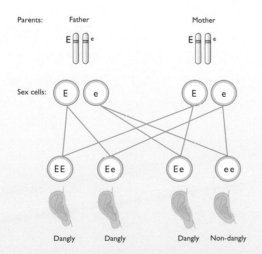

Parents: Father   Mother

            E ⬛⬛ e   E ⬛⬛ e

Sex cells: E   e   E   e

            EE   Ee   Ee   ee

            Dangly   Dangly   Dangly   Non-dangly

If both parents have one dangly allele and one non-dangly allele, their possible children are rather different.

Genetic diagrams can be used to predict the chance of a characteristic being passed on. The shape of our ear lobes seems to be passed on by one gene with just two different alleles. The allele that causes free hanging, dangly ear lobes is dominant so it is shown as "E". The allele for attached ear lobes is recessive and is shown as "e".

GENES WITH EVERYTHING

Whenever sexual reproduction takes place, genetics is at work, in living things from elephants to jellyfish, from giant redwood trees to moss. For hundreds of years people had no idea how information was passed from one generation to another. They knew that it happened, but they didn't know how. We owe our understanding to an Austrian monk called Gregor Mendel and his experiments with peas.

Gregor Mendel had ideas on genetics that were years ahead of his time.

SCIENCE PIONEERS Gregor Mendel

Gregor Mendel was born in 1822 in Heinzendorf in Austria, now part of the Czech Republic. He was incredibly clever – but very poor. In those days, the only way for a poor person to get an education was to join the church, so Mendel became a monk. He worked in the monastery gardens and became fascinated by the way things like colour and skin wrinkliness were inherited by the peas there. He decided to carry out some experiments, using pure strains of round peas, wrinkled peas, green peas, and yellow peas for his work. Mendel **cross-bred** the peas and made a careful note of the type of plants produced. He counted the different offspring carefully and found that characteristics were inherited in clear and predictable patterns. Mendel explained his results by suggesting there were separate units of inherited material. He realized some characteristics were dominant over others and that they never mixed together. To Mendel 150 years ago this was thrilling. The Abbot of the monastery was also interested in Mendel's ideas and built him a large greenhouse for carrying out his experiments.

Mendel kept records of everything he did, and analysed his results. Finally in 1866, Mendel published his findings. He explained some of the basic laws of genetics in a way we still use today. Sadly Mendel's genius fell on deaf ears. He was ahead of his time – no-one yet knew of the existence of genes or chromosomes, so people simply didn't understand his theories. He died eighteen years later with his ideas still ignored, but he was sure that he was right.

By 1900, people had seen chromosomes through a microscope. Three scientists, Hugo de Vries, Eric von Seysenegg, and Karl Correns, discovered Mendel's papers and duplicated his experiments. They gave Mendel the credit for what they observed. From then on, ideas about genetics developed quickly. It was suggested that Mendel's units of inheritance were carried on the chromosomes seen beneath the microscope, and the science of genetics as we know it today was born.

WORKING THINGS OUT

Smooth and wrinkled skin in peas is controlled by a single gene, in the same way as human dimples and ear lobe shapes. Features carried on a single gene like these are unusual – most of your characteristics are the result of several different genes working together. Things like the colour of your hair, skin, and eyes, the shape of your nose, and the length of your legs are the result of more than one gene working together and they are very difficult to predict. This is why the single gene features, although they are not very common, are so useful for understanding exactly how inheritance works.

To study genetics, scientists like to be able to choose the parents in a **genetic cross** and see lots of offspring as quickly as possible. This makes it difficult to use humans for genetic research. People usually produce only one baby at a time, which develops slowly over a period of years, and the average family has about two children in total – not enough to analyse using statistics. Also, people choose their own partners, so scientists cannot arrange a genetic experiment. All this means that human beings are not useful as laboratory specimens. Even Mendel's peas are not ideal, because it takes time and space for pea plants to grow, flower, and set seed. Ideally, scientists need a small organism that breeds quickly, has lots of offspring, takes up very little room, and is very cheap to keep. One **species** in particular makes a perfect specimen for genetic experiments – the fruit fly.

A large amount of what we know about genetics comes from work on this insect which most of us have never even noticed. The fruit fly *Drosophila melanogaster* was discovered about 100 years ago by the American scientist, Thomas Hunt Morgan. He noticed that the tiny fly came in different forms that were easy to recognize. Most fruit flies have red eyes. Thomas Hunt Morgan discovered fruit flies with white eyes and pink eyes. He also discovered flies with very short wings. He tried using them for genetic crosses, much like Gregor Mendel had done with his peas.

CLASSIC EXPERIMENT eye, eye!

Thomas Morgan's classic experiment was to cross fruit flies that had red eyes with those that had white eyes. He found that the first lot of offspring all had red eyes – but when the offspring bred together, some of the next lot of offspring had white eyes. He also noticed that males were much more likely to have white eyes than females were. Morgan was the first person to link genes, chromosomes, and the inheritance of sex into one neat package and his ideas all came from this first big experiment.

Drosophila melanogaster is very small, breeds very rapidly (it develops from egg to adult in just 12 days), it is cheap to keep and has lots of clear, easy to spot features which are inherited on single genes – ideal for geneticists to work with.

COUNTING THE CHROMOSOMES

Inheriting your granddad's red hair or your mum's nose doesn't really matter, but some of the things that can be passed from parents to their children are much more serious.

A normal person has 46 chromosomes, arranged in 23 pairs. But sometimes things go wrong and the number of chromosomes which gets passed on is more, or less, than usual. When this happens the developing **embryo** often dies and the mother suffers a **miscarriage**.

DOWN TO THE GENES

Sometimes babies are actually born with the wrong number of chromosomes. Most die in their first few months, but a specific group, children born with an extra copy of chromosome 21, do survive. This condition is called Down's syndrome. The extra chromosome causes problems as the brain and the body develop. Some of the problems can be quite mild, but others can be very serious; for example, the heart is often badly affected. People with Down's syndrome can often live to middle age leading fairly normal and fulfilling lives. The risk to the mother of having a baby with Down's syndrome increases as the mother gets older. There are special tests that can be carried out quite early in pregnancy that show if a baby has chromosome problems.

Children with Down's syndrome are often very loving. With lots of care and support they can make great progress and enjoy life to the full.

A DEADLY INHERITANCE

It seems amazing that the problems faced by someone with Down's syndrome are all caused by one extra copy of a chromosome. What is even harder to grasp is that other, far worse problems can be caused if you inherit the wrong allele of a single gene.

Huntington's disease is a rare but very serious genetic disease. It affects the **nervous system** when people reach middle age, and at the moment it is always fatal. It is inherited from one parent by a dominant allele of a gene. Unfortunately, because the disease does not usually show up until people are in their forties, most have already had children before they realize they have Huntington's disease. There is now a genetic test for people who have Huntington's disease in the family, but not everyone wants to take it. Some people would rather not know what their future holds.

If one parent has the allele for Huntington's disease, the chance of passing it on to any one child is 50 percent. The percentage chance is the same for each child. It is possible that all of the person's children could inherit the faulty allele. On the other hand, all of the children might inherit the healthy gene, or some of them might be affected but not others. The only way to know for sure is to have a genetic test for Huntington's disease.

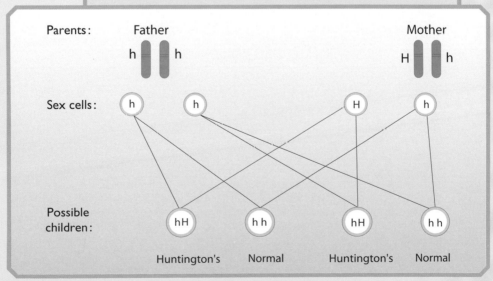

MORE GENETIC PROBLEMS

Huntington's disease is rare, and also quite unusual because the allele that causes the disease is a dominant one. Many other genetic diseases are caused by recessive alleles.

With the disease **cystic fibrosis**, the body makes a lot of thick, sticky mucus which stops the lungs, the gut, and the reproductive system from working properly. Cystic fibrosis is caused by a recessive allele, so it has to be inherited from both parents. They will both have a copy of the dominant healthy allele so their bodies work normally. But they carry the cystic fibrosis allele – and usually have no idea it is there. They are known as **carriers**. In the past, children born with cystic fibrosis did not usually live very long, but modern medicines and treatment mean that many people affected by cystic fibrosis now live full, active lives, and their life expectancy is getting longer all the time. With a lot of determination, some young people affected by cystic fibrosis achieve more than people with healthy genes.

Not every problem you can inherit from your parents is as serious as Huntington's disease or cystic fibrosis. Inherited problems such as colour-blindness may cause inconvenience in everyday life, but are not as serious or life-threatening. And as you will see on page 37, some genetic diseases can carry unexpected benefits.

THE INFLUENCE OF THE GENES

It is not just genetic diseases that are affected by your genetic makeup. It seems that many other diseases are influenced by your genes as well. Some people have a combination of different alleles that make them much more likely to develop **cancer** than others. Heart disease also seems to run in families, and scientists are discovering genetic reasons for more and more health problems. But these genetic tendencies are very different to the true genetic diseases. Often lifestyle plays an equally important part in deciding your health.

RECENT DEVELOPMENTS cancer treatment

Research teams in the UK have found some interesting genetic links in the struggle to understand and treat cancer. For example:

- Dr Michelle Guy and her team at St George's Hospital in London have found that some alleles of a gene controlling the use of vitamin D in the body seem to be linked to a higher risk of developing breast cancer.
- In Scotland, Dr Andrew Schofield and his team have found that if a woman has certain alleles of a gene called p27, her cancer cells will not be killed by a commonly used anti-cancer drug. This means that the scientists can work out alternative treatments.

Many people run the London marathon to raise money for charities researching genetic diseases. In 2005, 1,650 people ran for Cancer Research UK and they raised over £1 million.

Variety packs

When animals or plants reproduce asexually, the offspring are identical to their parents, but sexual reproduction gives us variety. In sexual reproduction, genetic information from two different parents joins together, and the effect of all the different dominant and recessive alleles is lots of variety in the offspring. Look at a litter of kittens, a shoal of fish, or even your own brothers and sisters, and the differences are plain to see. But all of this variety is not just the result of different combinations of genes. There is another force at work – **mutation**.

MUTATION

Mutation is often seen in films and other stories as something dramatic, causing huge, scary, and revolting changes. The facts about mutation are much less exciting, but much more important. All the living things on the Earth are the result of millions of mutations which have taken place since the earliest days of life on the planet.

A mutation is just a change in a gene – the appearance of a new allele. They come about when tiny changes are made in the long strands of DNA. Mutations often happen naturally when

Mutation of the DNA in fiction is fun – but often not very realistic!

mistakes are made in copying the DNA for new cells in your body. If mutations take place in your sex cells, they can be passed on to the next generation.

Some things increase the chances of mutations happening. If your cells get lots of **ionising radiation**, such as **ultraviolet light** from the Sun, X-rays, or radiation from radioactive substances, mutations in the DNA are more likely.

MUTATIONS – GOOD OR BAD?

Some mutations are certainly bad for us – for example, some chemicals in cigarette smoke (known as **carcinogens**) cause mutations in the cells of the lungs, which can lead to lung cancer. Mutations in the sex cells can cause the **foetus** to die at an early stage of pregnancy, or they can lead to serious genetic diseases. But most mutations do not have any effect at all. They take place in parts of the DNA that do not affect how an organism looks or works, so there is no noticeable difference.

However, occasionally a mutation can produce a change for the good, something that helps an animal or plant to survive more easily. Those mutations survive as the organisms reproduce, adding to the variety of life. If the mutation is really useful, it will become more and more common.

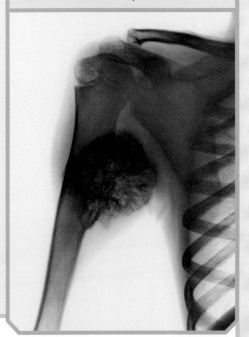

A mutation in a single cell can trigger the growth of a huge tumour, like this one which can be seen on a person's arm.

Did you know..?

When a mutation takes place in a body cell, normal cell growth can get out of control. The cells divide very quickly to form a tumour, and bits may break off and invade other parts of the body. This uncontrolled growth is what we call cancer.

VARIETY — THE SPICE OF LIFE

It is often very easy to spot the differences between species – not many people would mistake a carrot for a cat! But the variation between some species is so small that most of us would not be able to tell them apart. Great or small, the variation between species is the result of genetic differences.

However, there can also be a lot of variety between members of the same species. Although much of this is down to the genes, the situation in which an organism lives is also a factor.

Height can vary tremendously in the human species. It is affected both by the genes and the environment in which the person grows up.

You can find the same species of tree as a tall and graceful specimen in a public park and as a low, stunted bush growing on the face of a cliff. Wild gerbils are fast moving and lean, but a pampered and overfed pet gerbil might move slowly and weigh twice as much or more. Variety that is the result of genetic differences can be passed on from parents to their offspring, but variety that is the result of lifestyle or **habitat** differences cannot be passed on.

NATURAL SELECTION

Living things are always in competition with each other. Plants compete for light, water, and minerals from the soil. Animals compete for food, water, partners for reproduction, and for space to live. When a mutation causes a change that is useful to an organism, that animal or plant gets an advantage in the competition both against other species and against other members of its own species. So individuals with the new characteristic are more likely to survive and breed – and this is known as **natural selection**.

Charles Darwin was the first person who explained natural selection as the "survival of the fittest", but how does it work? Reproduction is one of the great driving forces of life but it is very wasteful. There are always more offspring than the environment can support. Think of all the frogspawn that appears every spring, or the seeds on a dandelion head. If all those offspring survived, the world would be overrun with frogs and dandelions. So the offspring with the genes best suited to surviving in the place where they live will be most likely to stay alive and breed successfully. This is natural selection.

Did you know..?

The tallest human being ever recorded was an American man called Robert Pershing Wadlow. He grew to the incredible height of 2.72 metres (8 feet 11.1 inches). At the other extreme, Douglas and Claudia Maistre of Brazil are the shortest married couple on record. They are 90 centimetres (35 inches) and 93 centimetres (36 inches) tall!

NATURAL SELECTION IN ACTION

There are millions of different species of living organisms in the world today. They have all come about as a result of natural selection. Chance genetic mutations have changed populations of animals and plants to make them better able to survive in the place where they live.

KEY EXPERIMENT
The oysters of Malpeque Bay

In 1915, the oyster fishermen of Malpeque Bay in Canada noticed a few diseased oysters with pus-filled blisters. By 1922, the oyster beds had been almost wiped out by this so-called Malpeque disease. However, by chance, a small number of the millions of baby oysters produced each year carried an unusual allele that gave them resistance to Malpeque disease. Not surprisingly, these were almost the only oysters that managed to survive and carry on breeding. By 1925, the oyster beds were recovering and in 1940 there were as many oysters as before the disease, but now they were almost all resistant to the disease.

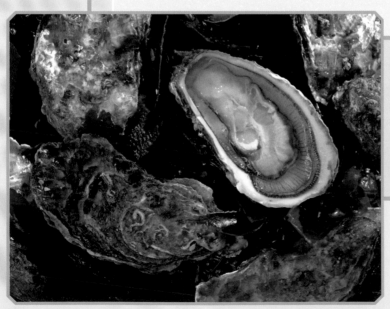

Oysters have genes too. Malpeque Bay was a case of natural selection in action. A chance mutation at some point produced an allele that saved the oyster beds from a deadly disease.

SCIENCE PIONEERS Charles Darwin

In 1831, a 22-year-old Charles Darwin set out on his voyage on *HMS Beagle*. Darwin intended to study mainly geology, but as the voyage progressed, he became more and more excited by his collection of plants and animals. By the time the expedition reached the Galapagos Islands, Darwin was amazed at the variety of species he was finding and the way they differed from island to island. He found strong similarities between the giant tortoises and the mocking birds on the different islands, yet he also observed clear differences, which meant that each animal had **adapted** to make the most of the local conditions.

By the end of the five-year voyage, Darwin had collected so many specimens, and made so many drawings and notes, that he had enough work to last him for years. In fact, it was over 20 years later, after studying all his specimens closely and working on his pigeon breeding programmes, that Darwin finally published his now-famous book, called *On the Origin of Species by Means of Natural Selection*.

Charles Darwin's ideas are still an important part of the way we look at biology.

Did you know..?

In 1832, Darwin's party in South America cooked and ate a bird they had found. During the meal Darwin realized they were eating a completely new type of bird. He sent the remains of the meal back to England to be identified. It was later named *Rhea darwinii* in his honour.

MORE NATURAL SELECTION

Natural selection takes place between the members of the same species, and it also helps to form separate species. One clear example of natural selection in action is with the Arctic hare. There are several different alleles for coat colour in these hares, but the most common ones are brown and white. For the Arctic hares that live in areas with a lot of snow, a white coat is a real advantage. A pure white hare sitting in the snow is much less likely to be noticed by a fox or owl than a light brown hare, so it is much more likely to survive long enough to breed and pass on the alleles for a white coat. The great majority of hares found in the northern areas of the Arctic are white. On the other hand, hares that live a bit further south make their homes in coniferous forests. The surrounding landscape is brown rather than white. Here the white hares stand out and are easy targets, while the hares with the alleles for brown fur fade into the background and live to breed and pass on their successful genes.

The white fur that gives the Arctic hare such good camouflage in the snow would cause them problems if they live somewhere warmer.

SUPERBUGS – A SUCCESS STORY FOR BACTERIA

The selection of a genetic trait that is really helpful to one type of organism can cause all sorts of difficulties for others. There are more bacteria than any other living organism on the planet. They are **micro-organisms**, and although most of them are harmless or even useful to human beings, some of them can cause diseases. **Antibiotics** are medicines that can kill the bacteria to prevent them from causing disease.

Usually, when an antibiotic is used, the bacteria are all wiped out. Sometimes, however, there are a few that carry a mutant gene making them resistant to the antibiotic, so they are not killed by it. These few that survive then reproduce and pass on their antibiotic-resistant genes. Fortunately, there are a number of different antibiotics, so if bacteria are not killed by one, another can be tried. But bacteria can become resistant time after time. There are now some bacteria known as "superbugs" that seem to be resistant to all antibiotics. Superbugs can be a real threat, and many people have died. Natural selection has done wonders for the bacteria, but created a huge problem for us.

The making of a superbug. This is why it is so important to finish a course of antibiotics – the most resistant bacteria will be the last ones to be killed. If you take all of your medicine, you help to prevent superbugs developing.

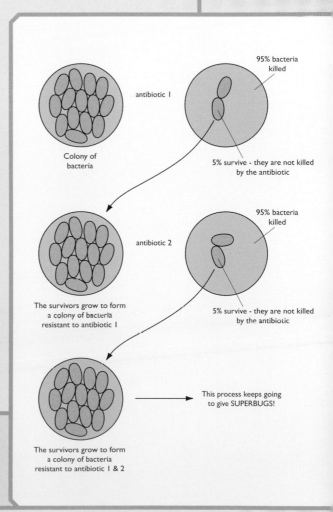

Colony of bacteria

antibiotic 1

95% bacteria killed

5% survive - they are not killed by the antibiotic

The survivors grow to form a colony of bacteria resistant to antibiotic 1

antibiotic 2

95% bacteria killed

5% survive - they are not killed by the antibiotic

The survivors grow to form a colony of bacteria resistant to antibiotic 1 & 2

This process keeps going to give SUPERBUGS!

BAD MUTATION, GOOD MUTATION?

Sometimes a genetic mutation that can cause big problems in some situations can be useful in other ways. The variety of life arises from the genes and is affected not only by changes in the DNA (mutation), but also by changes in the environment that make different alleles more or less favourable. You can see this very clearly when you look at the case study of **sickle cell anaemia**.

This picture clearly shows the difference between normal disc-shaped red blood cells and the strange shaped red blood cells caused by sickle cell anaemia.

CASE STUDY blood cells going wrong

Normal red blood cells carry oxygen around the body. They are disc shaped with a dimple on each side, which is ideal for carrying oxygen. The shape is controlled by a single gene. If there is a mutation in both of the alleles, the shape of the red blood cells changes to a thin, shrunken sickle shape. This shape does not allow the cells to carry oxygen properly and they get stuck in the smallest blood vessels.

Without treatment, children affected by sickle cell anaemia usually die young, but **blood transfusions** can help to keep them well. Sickle cell anaemia is a major problem, but in some cases the sickle cell allele can in fact be life-saving for the people who carry it. Someone with normal alleles for red blood cell shape has normal blood, while someone with two mutant alleles has sickle cell anaemia. But people (known as carriers) who have one normal allele and one mutant allele have an amazing advantage. Although they suffer slightly from sickle cell anaemia, they are resistant to malaria!

Malaria is an infectious disease that infects the red blood cells. It is spread by the bite of mosquitoes. Malaria causes illness to millions of people across the world every year, and sometimes even death. The areas of the world where malaria is particularly common (including large parts of Africa) are also the places most affected by sickle cell anaemia. The sickle cell anaemia allele saves the lives of hundreds of thousands of people who are just carriers of the disease. While people live in areas where malaria is common, the sickle cell allele has also become common and is a big advantage. But where malaria is not a problem, the advantages of carrying the allele are lost and only the difficulties remain.

Selective breeding

Natural selection has been around as long as life on Earth. The animals or plants that are best suited to living in a particular place are the most likely to survive and breed. If conditions change, the organisms that cope best with the new environment will survive. This is how gradual changes in the living world take place.

From the earliest recorded times, people have tried to speed up the slow process of natural selection. They have chosen animals and plants with features they particularly like and bred from them. Farmers bred from the plants that produced the biggest grain to get plants that all gave more grain, or the animals that produced the most milk to get animals that all produced a lot of milk. This is known as selective breeding.

SELECTIVE BREEDING SUCCESS

The world we live in is full of the results of selective breeding. If you have a dog or a cat, it will not look much like the wolf or wildcat that is its distant ancestor. Not only have people selected animals with good natures to act as pets, they have also selected for all sorts of different shaped ears, noses and tails, sizes, colours, and types of fur. There is a dog or cat to suit almost everyone!

The flowers in a flower shop do not look much like wild flowers either. They have been bred over the years to have bigger flowers, more colourful flowers, lots of flowers, a lovely scent, and to be resistant to moulds and viruses. The effects of selective breeding can also be seen very clearly when looking at horses.

Horses range from enormous Shire horses and draught horses, through the lean and super-fast Arabs and thoroughbreds, to the tough native breeds such as the American mustang, the Australian Brumby, and the tiny Shetland pony, bred to work hard and eat very little in the harsh conditions of the Shetland Isles off the coast of Scotland.

Most selective breeding has been carried out to produce more food. The range of cereals now available to farmers around the world do not look much like the wild grasses they originally came from, any more than dairy cattle look like the wild version. So how are these dramatic changes brought about?

People have selected all sorts of different characteristics in dogs – small to fit down a rabbit burrow, fierce and strong to warn off intruders, fast for hunting and racing, or with lots of hair to keep the dog warm.

Did you know..?

One of the most extreme examples of selective breeding in the dog world is the American hairless terrier which has almost no hair at all. It is the result of selective breeding to produce an ideal pet for anyone who is allergic to dog hair!

FROM GRASS TO GRAIN...

Getting enough food to eat has always been vital for humans. Almost everywhere in the world, people rely on some kind of cereal to provide a major part of their diet. The cereal grown varies around the world, from the rice in the paddy fields of Asia to the sorghum that grows in very hot, dry conditions, and the maize that thrives on sunshine and water. In every case, cereal plants have been selected over generations for the characteristics that now make them successful.

In the same way, dairy cows now produce many litres of milk every day, far more than the original wild cattle who would produce enough for their own calf and little more.

Mass production of milk was made possible by careful selective breeding.

HOW DOES SELECTIVE BREEDING WORK?

We can change animals and plants by artificially selecting which ones are allowed to breed. With animals like cattle and sheep, it is easy to see how this can be done. For example, with cattle, you choose a bull with characteristics you want, such as size or good-nature, and breed him with cows that also have the same feature. You then choose the best offspring and use them to breed the next generation, and so on.

The same idea is used for plants. For example, you may choose two wheat plants that you want to breed. The pollen must be transferred from one flower to another very carefully, well away from any other similar plants to make sure you get the cross you want. After many years, perhaps indeed centuries, the results of selective breeding are dramatic.

SCIENCE PIONEERS Charles Darwin and his pigeons

The animals and plants that Charles Darwin saw during his five-year voyage on *HMS Beagle* helped him to develop his ideas on natural selection. However, he also used what he knew about artificial selection as he worked on his theory. Darwin was a very keen breeder of fancy pigeons, selecting birds carefully to produce offspring with the features he wanted. Working with his birds helped him make a link between what people had been doing for years in artificial selection and what was happening in the natural world.

Did you know..?

Selective breeding of cattle has increased the average milk yield for a cow to about 12,990 litres (22,860 pints) of milk a year. One record-breaking cow produced 22,260 litres (39,170 pints) in just 305 days!

BACK TO MENDEL: SELECTIVE BREEDING

Selective breeding can bring about all sorts of changes in plants and animals but it is not always as easy as it sounds. Looking at a simple example can help to explain the problems. Imagine you have a plant that has red flowers, and you want to produce lots of red-flowered plant seeds to sell. It seems that the obvious plan would be to choose another red flower, cross-pollinate them, and wait for lots of new plants with red flowers.

The problem is that the allele for red flowers is dominant over the allele for white flowers. This means it is impossible to tell just by looking at the flowers whether the plant is homozygous (sometimes known as true breeding) or heterozygous for the red flower allele. Unless you use two plants which are homozygous for the characteristic you want – in this case, red flowers – then some of the offspring might appear with white flowers. That could lead to some very unhappy customers!

BEWARE THE F1 HYBRID!

Seed packets look wonderful but they can be quite expensive. Lots of gardeners like to try and collect seeds from the plants they grow to plant again the next spring. But often the seeds sold today are F1 hybrids, and the seeds from those plants are no use at all. F1 hybrids are the result of a cross between two parents which both bring very good characteristics to the offspring. In the case of bean plants, perhaps one parent is true-breeding for a dominant allele giving long seed pods and the other is true breeding for a dominant allele giving large, tender beans. The F1 plants will be excellent – in our example, carrying long pods filled with large tender beans – but they will not necessarily be true breeding. They may also carry some recessive genes from their parents. This means that plants grown from the seeds produced by the F1 plants might be very good, but they might be awful – you might get short pods full of tiny, hard beans, for example!

Seed packets tempt us with the beautiful flowers and vegetables we might be able to grow – but beware the F1 hybrids if you want to plant your own seed next year!

Did you know..?

F1 hybrids cause a lot of problems for farmers in developing countries. In the past, they would save seeds from one year to another. But many of the new cereal seeds that give them the better crops they need are F1 hybrids. This means they have to buy new seeds each year from countries such as the US and the UK.

SUPERDADS...

In the past, if someone wanted to improve their animals they had to choose their breeding stock from other local farmers. Nowadays, a farmer in the United States might decide they want to breed their cattle with a bull from the UK – and it can be done without the bull ever leaving home. **Artificial insemination (AI)** has made it possible for farmers to use bulls from anywhere in the world to **fertilize** their cows. **Semen** is collected from the top breeding stock at regular intervals and frozen in liquid nitrogen to store it. The small frozen tubes can be transported all over the world. When it is thawed out, one batch of semen can be used to inseminate a number of cows – and the offspring that result will have the genes of one of a small group of top class bulls. The same can be done with pigs.

...AND FLYING COWS!

Another modern aspect of selective breeding is the arrival of flying cows – and pigs, and sheep as well! Not the real thing, of course, but embryo animals flying all around the world. Female animals only produce a

Selective breeding using animals is no longer just a local business – frozen semen samples mean champion animals such as this ram can father huge numbers of offspring around the world.

small number of eggs, but **hormones** can make them produce more than usual. Once they have been mated with top quality males, the tiny embryos are removed. They can be placed in the **uterus** of another animal (like a rabbit) for a short time. The rabbits, containing their valuable cattle embryos, can be flown easily and cheaply anywhere in the world. Each embryo is then transferred from the rabbit back into the uterus of a cow, which later gives birth to a very special calf.

In some cases the embryos are also **cloned** before they are put into the rabbits. This makes a number of identical embryos, and produces even more highly-selected offspring growing up a long way from home. These techniques have been used to help improve the breeding stock of cattle in areas of the world where more milk and meat is desperately needed, in just the same way that selective breeding has also been used to increase crop yields where they are most needed.

RECENT DEVELOPMENTS red bluebonnets!

In 1984, Dr Jerry Parsons, an American agricultural scientist, began a project to grow red bluebonnets (normally a blue flower). He found a patch of rare pink bluebonnets in Bexar County, Texas and when they had reproduced, he used only the plants with the darkest pink flowers to breed again. A few generations later he had maroon bluebonnets – he could never get rid of the original blue completely, though.

It took 15 years of selective breeding for Jerry Parsons to change the colour of these flowers!

Selecting the future

As you have seen, selective breeding has been around for thousands of years. However, things really started to speed up during the 20th century. Until then, selective breeding had depended on the appearance of natural mutations to introduce a new and desirable feature. But in the late 1920s, scientists discovered they could greatly increase the numbers of mutations that cropped up by exposing plants to X-rays and certain chemicals.

This new technique, known as mutation breeding, really took off after the Second World War when all sorts of different ionising radiation had been discovered and tried on plants. Valuable strains of wheat, barley, rice, potatoes, soybeans, and onions are just some of the 2,252 types of plants that have been produced in this way. Almost half of them have been developed in the last 15–20 years. Although the technology of mutation breeding has been overtaken by some other techniques – particularly **genetic engineering**, which is a very new and expensive process – it is still important in the artificial selection of plants.

Rather than just being used to increase the amount of food we can produce, this method is also used for our pleasure. People want ornamental plants for their gardens, but they do not always want to pay huge prices for them. So mutation breeding is still popular with the people who breed ornamental plants around the world, and helps them to make our gardens full of interesting flowers and colours.

THE IMPORTANCE OF CLONING

Almost everything covered in this book so far has been about variety – about the mixing of the genes during sexual reproduction, how mutation leads to variation, and how variety is vital both for natural and artificial selection. But one form of reproduction, used more and more in artificial selection, makes for no variety at all. This is called **cloning**.

What image does the word cloning conjure up in your mind? *Jurassic Park* with its laboratories full of cloned dinosaur embryos, *Star Wars: Attack of the Clones*, or even cloned human beings? Cloning has had a lot of sensational media coverage, but the facts are rather more down to earth.

A clone is a group of cells or organisms that are genetically identical and have all been produced from the same original cell. All of the bacteria, plants, and animals that have been produced as a result of asexual reproduction are also clones. In fact, people have been cloning plants for centuries by taking **cuttings**. The big change over the last 50 years is that we have developed the ability to produce clones artificially. Not only that, but we can produce animal clones as well as plant clones.

If you know any identical twins, you know some clones!

Did you know..?

Identical twins are natural clones of each other, where one fertilized egg cell has split completely to give two genetically identical embryos which then grow into genetically identical people.

CLONING PLANTS...

Although cloning itself does not produce any variety, it is very important in the artificial selection of varieties. Taking cuttings, splitting clumps, and dividing bulbs are all ways of cloning plants. But now we can make clones from a single cell, and this is often used to produce lots of identical copies of a new variety of plant that is the result of mutation breeding.

Using a special mixture of hormones and a jelly full of nutrients, a single cell from a plant can be used to produce a great big mass of identical cells. If each of these cells is separated and given a mixture of hormones and nutrients, it will grow into a tiny new plantlet. This kind of plant cloning makes it possible to reproduce new varieties of plants very quickly.

Cloned cells can give rise to thousands of identical plants. These tiny plantlets are from one plant cell, so they have the same genes.

...AND CLONING ANIMALS

Cloning plants is one thing, cloning animals is quite another. Animals, such as sheep or cats, are extremely complicated. They have lots of different types of cells which are specialized for doing particular jobs, and as a result many of their genes are switched off. But if you want to make a whole new organism, all of the genes need to be switched on again.

Cloning animal embryos has become quite common. Scientists have simply taken the natural process by which identical twins are formed a few steps further, allowing the embryos of prize livestock to become small balls of cells. They then separate the cells and let each one develop into a new embryo.

But true cloning of mammals, without any sexual reproduction involved at all, only really began in 1996, when a team of scientists in Edinburgh produced Dolly the sheep.

KEY EXPERIMENT
the cloning of Dolly the sheep

Professor Ian Wilmut and his team at the Roslin Institute worked for years on the problems of cloning sheep. Dolly was produced in 1996 by a complicated process and she was the only one of hundreds of attempts that produced a live lamb.

The nucleus was removed from a cell taken from Dolly's biological mother (the **donor** sheep). At the same time, the nucleus was removed from the mature egg of another sheep. The nucleus from the donor cell was put into the empty egg and given a mild electric shock, which made it begin dividing to form an embryo. The embryo was then transferred into the uterus of another sheep where it could develop. When Dolly was born, she was genetically identical to the ewe that gave the original cell, not the ewe that gave birth to her!

Dolly the sheep was the first mammal cloned from an adult cell.

GENETIC ENGINEERING

In the past, it took years and years of careful selective breeding to change something in an animal or plant. Now it can be done very quickly in a process know as genetic engineering. In selective breeding you can only select for genes that are already there, but using genetic engineering you can put genes from one organism into a completely different one. For example, you can put genes from a person into a sheep, or even into a bacterium!

WHAT IS GENETIC ENGINEERING?

Genetic engineering involves changing the genetic material of an organism. It usually involves taking a gene from one organism and moving it into the genetic material of a completely different organism. Genetic engineering often involves the use of bacteria

Genetic engineering looks easy in a diagram like this. In real life, it is much more difficult!

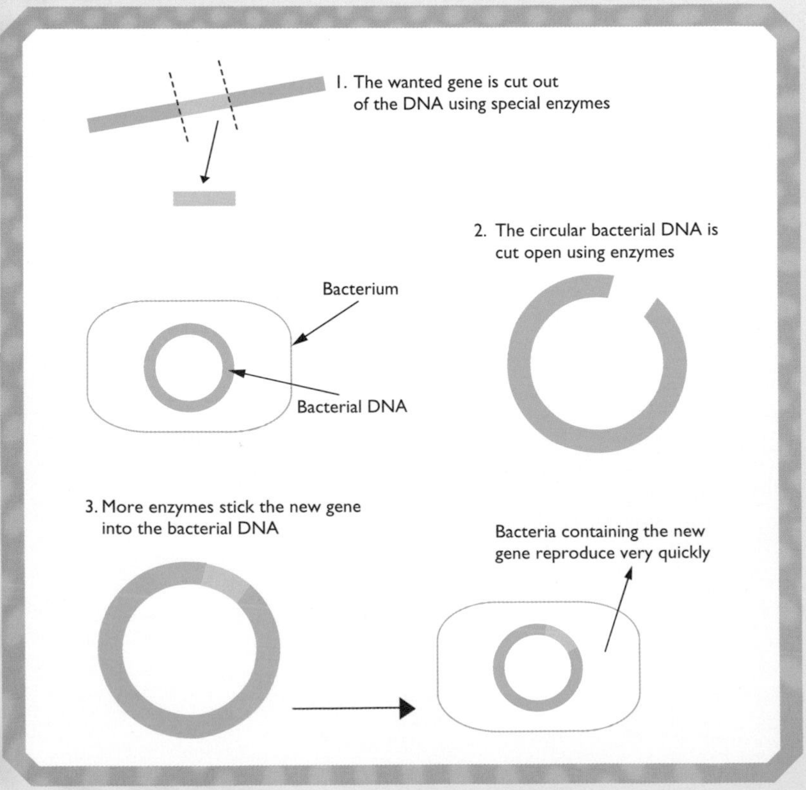

1. The wanted gene is cut out of the DNA using special enzymes

2. The circular bacterial DNA is cut open using enzymes

Bacterium

Bacterial DNA

3. More enzymes stick the new gene into the bacterial DNA

Bacteria containing the new gene reproduce very quickly

or viruses. For example, a gene from a person can be taken out of a chromosome using special enzymes. At the same time, the DNA in a bacterium is split open. The human gene is stuck into the bacterial DNA using more special enzymes. Once it is in place, the bacterium will "read" the human gene along with the rest of its own genome – and so it will begin to make a human protein. Once an organism has been "engineered", it is usually cloned to make lots of copies of the new DNA.

WHAT ARE THE BENEFITS?

One big advantage of genetic engineering is that changes to an animal, bacterium, or plant that would have taken many years to introduce, or would have simply been impossible, can now happen in a very short time. Another is that organisms can become much more useful to people. For example, bacteria have been engineered to make a number of human proteins that are used as medicine – the best known is probably **insulin**.

CASE STUDY human insulin from bacteria

Many people affected by **diabetes** need to inject themselves with the hormone insulin several times a day to control their blood sugar levels. For many years, the insulin they needed came from the **pancreases** of cattle, sheep, and pigs slaughtered for food. There were lots of problems with this. The supply of insulin depended on how much meat people ate (to provide the pancreases), the concentration of the insulin could vary, and some people were allergic to the animal insulin.

Then in 1982, human insulin made by bacteria became available. Scientists had genetically engineered bacteria to contain the gene for human insulin. When the bacteria grew, they produced pure human insulin. Now people with diabetes have a reliable, pure source of the insulin they need to live. The impact of genetic engineering on the treatment of diabetes has been tremendous.

GENETIC ENGINEERING IN MEDICINE

There are two main areas in which genetically modified organisms are being used – in medicine and in agriculture. The production of insulin by bacteria is just one example of the way in which genetic engineering has made medical advances possible. However, there is a limit to the type of proteins that bacteria can make, and so genetic engineering has been extended to much bigger animals, including sheep and cattle. It is much more difficult to transfer new DNA into mammals – not least because they have a lot more DNA than bacteria. But **transgenic** mammals (animals that contain a gene transferred from another species) have been developed to solve the problem. In 1991, Tracy the sheep was the first transgenic sheep. These animals produce human proteins in their milk, which makes it very easy to collect the medicine they have made. Some of the most exciting transgenic animals so far have been sheep that make human blood-clotting proteins. These are life-saving to people with rare blood-clotting diseases who would otherwise bleed to death.

GENE THERAPY

Many people hope that genetic engineering might be used to cure some of the dreadful genetic diseases that some people are born with. The idea is that a healthy gene might be put into some of the cells of the person with the disease, replacing the faulty mutant gene. If those healthy cells take over from the damaged ones, the body should be able to start working normally. **Gene therapy** has been tried for some diseases, but so far it has not been as successful as scientists hoped. But it is still a very new science and these are early days.

Children affected by the very rare genetic disease SCID (Severe Combined Immunodeficiency) have immune systems that do not work. Several children have been treated successfully using genetic engineering and have been able to live normal lives. Sadly, some of the children then developed cancers, so scientists are making sure the method is safe before more children are treated.

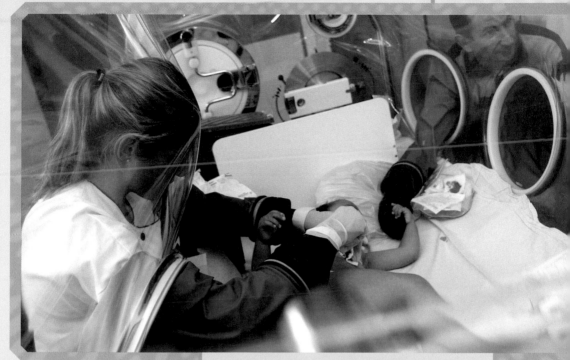

Children with SCID have to be looked after in a safe, clean environment. They may catch life-threatening infections because their immune system does not protect them.

RECENT DEVELOPMENTS the battle against cystic fibrosis

Cystic fibrosis is caused by two faulty recessive alleles. Scientists have been trying to develop a genetic engineering treatment for cystic fibrosis to place a copy of the dominant healthy allele into the affected cells. So far scientists are concentrating on the cells of the lungs, as they are easiest to reach and also cause most of the serious problems. There have been some difficulties in placing the healthy gene into the cells and it has taken much longer than people hoped. But scientists have kept working on the problem and new methods of getting healthy genes into the cells are giving much better results.

GENETIC ENGINEERING ON THE FARM

Genetically modified (GM) plants are much more common than engineered animals. This is partly because it is much easier to work with plants than with animals, and partly because there are so many changes to plants that are worth a lot of money.

So what sort of GM crops might be growing at a farm near you? Plants including cotton, potatoes, and tomatoes have been given a gene that makes them produce insecticide in their tissues, so if a pest eats the plant, it is poisoned. Potatoes and oil seed rape have been engineered to be resistant to herbicides (weed-killers), while oil-producing plants, such as soya beans and oil seed rape, have been genetically modified to produce more oil. These oil-producing plants have also been modified to produce oils that are useful for particular jobs, such as making cosmetics or healthier foods.

PREVENTING LOSS

More than half of the food that is grown around the world is lost during storage and transport because it goes mouldy, becomes over-ripe, or gets damaged in some way. Simply solving this problem could help prevent people starving in many countries of the world.

Genetic modification is being used to produce fruit that takes much longer to ripen, and so is much more likely to survive being stored and transported. Of course, GM fruits are also able to last a lot longer on supermarket shelves, which makes a bigger profit for the sellers.

There seem to be many advantages to using genetically engineered plants. However, they are not yet widely used around the world. In fact, there is so much concern about possible problems that many countries will not allow GM food to be grown or sold at all, or to be sold only if the food is clearly labelled GM so people have a choice about what they are eating.

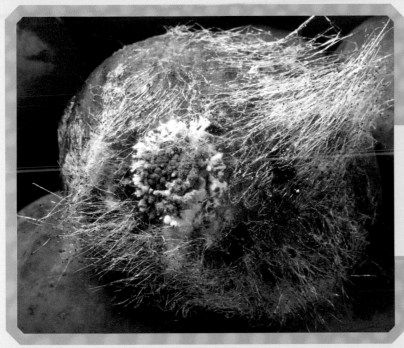

In the United States and Europe, around 40 percent of the food can be lost during storage and transport. In some parts of the world this can be up to 80 percent!

CASE STUDY the GM dilemma

People have several concerns about GM food. Locally, people worry that eating GM food will affect human health or that the GM plants will breed with wild plants and cause problems like so-called superweeds. On a world scale, people worry that GM plants will cost too much for developing countries to afford. What is more, some of the modified plants have a "terminator" gene (which means they do not produce fertile seeds), so farmers in poor countries will not be able to harvest their own seed. They will have to buy new seed from the rich countries every year, which they cannot afford to do.

Many scientists are working hard to develop more GM crops and to convince people that they are safe. Other scientists are convinced there are more problems to come. Most of us just have to wait and see who is right.

MOVING FORWARD – BUT NOT TOO FAST!

Genetic engineering is a very exciting and fast moving area of science, but there are also lots of voices warning about possible dangers that may lay ahead.

GENETICALLY MODIFIED PEOPLE?

The biggest hopes for genetic engineering are linked to human medicine – but so are the biggest fears. So far, gene therapy has been aimed at body cells. But there is another possibility. **Conception** can now take place outside the human body in the process known as *in vitro* fertilization (IVF, commonly known as test-tube babies). Already those very early embryos are sometimes screened so that couples affected by a genetic disease know they will have a healthy baby.

In future it may be possible to go one step further and genetically modify the very early embryos. This would get rid of the faulty gene completely – the baby would be born healthy and the gene would not be passed on again. Changing the genes of an early embryo is known as **germ line therapy**.

There are many **ethical** problems to be considered before germ line therapy is tried. The technique might affect the way an embryo develops – it would be terrible to solve one problem but make another. Most people see preventing genetic disease as a good thing, but many fear that once it is possible to change the genetic makeup of an embryo to prevent disease, some people would be prepared to pay enormous amounts of money to make sure that their embryo was intelligent, or tall, or beautiful, and so on. So far, germ line therapy is banned around the world.

SCIENCE PIONEERS Dr Charles Strom

In the year 2000, Dr Charles Strom, working in the US, selected an embryo to implant into the womb of Lisa Nash. Adam, the baby born 9 months later, was chosen out of 16 embryos. Before he was born his parents knew he was free of Fanconi anaemia, the fatal genetic disease which was affecting his 6-year-old sister Molly. They also knew that Adam's cells were a perfect match for Molly so that cells from his umbilical cord could be used to help save his sister's life.

Charles Strom knew that this was a big step – selecting an embryo not just for its own health but also to save another child. He and his team felt that it was a good use of technology already in place. Some people disagree, partly because the other embryos that did not match Molly were destroyed. Cutting edge technology such as genetic modification and selection continue to raise many ethical issues.

Adam was the first baby to be born who had been selected partly to help save his sister's life – but he has not been the last! The work was started in the US in 2000, and at the end of 2001 it also became legal in the UK to use this technique to help families with serious genetic diseases.

A brave new world?

Much of our new knowledge about inheritance and selection raises ethical questions and **moral** issues which have to be dealt with both by individuals and by society as a whole – there is a lot to think about.

BANANA VACCINES

Vaccinations protect people all over the world from dreadful diseases. However, there are many countries where keeping vaccines cold and giving injections to hundreds of thousands of people is really difficult. Scientists are working on a genetically modified banana that will act as a vaccine against hepatitis B, a liver disease that affects more than 2 billion people worldwide. Bananas grow easily in most of the countries that need the vaccine, they are eaten raw, and they are very popular with children.

In the future, many vaccines and other medicines may be given to people and animals in easy-to-eat fruit and vegetables – a pleasant change from the days of hypodermic syringes!

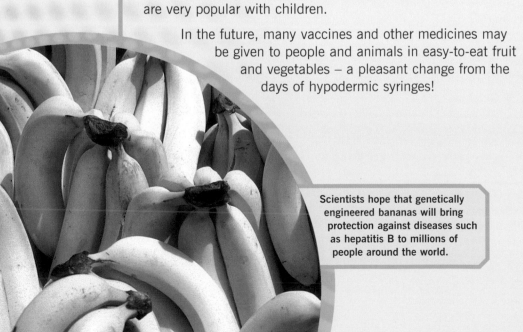

Scientists hope that genetically engineered bananas will bring protection against diseases such as hepatitis B to millions of people around the world.

XENOTRANSPLANTATION – PUSHING THE BOUNDARIES

Every year, thousands of people die because they need an organ transplant but there are not enough donor organs to go round. On top of this, our immune systems recognize a new organ as foreign tissue and try to reject and destroy it. If you have an organ transplant, unless it comes from your identical twin, you will need anti-rejection medication for the rest of your life.

Genetic engineering brings hope of a possible solution to both problems. Scientists have produced genetically-engineered pigs which have been modified so that the cells of their hearts no longer trigger an immune response when transferred into human beings. If pig hearts – or lungs, or kidneys – could be safely transplanted into humans without any risk of rejection, then thousands of lives could be saved each year. Scientists have been working on this technique, called **xenotransplantation**, for a number of years now, but the use of modified pigs seems to have caused several problems. The biggest risk is that pig diseases will cross into people and cause great damage. Also, for many people there are ethical issues about the use of animals in this way. For the time being at least, this is one area where genetic engineering is raising problems, not finding solutions.

When we look at inheritance and selection in the 21st century, we are looking at sciences that are still very new but that have enormous potential to benefit us all. We just have a lot of learning and understanding still to do!

Further resources

MORE BOOKS TO READ

George, Linda, *Gene Therapy* (Blackbirch Press, 2003)

Gillie, Oliver, *Need to Know: Sickle-Cell Disorder* (Heinemann Library, 2004)

Stockley, Corinne, *The Usborne Illustrated Dictionary of Biology* (Usborne Publishing, 2005)

Nature Encyclopedia (Dorling Kindersley, 1998)

USING THE INTERNET

Explore the Internet to find out more about inheritance and selection. You can use a search engine, such as www.yahooligans.com or www.google.com, and type in keywords such as *genes*, *DNA*, *chromosomes*, *humane genome project*, or *GM crops*.

These search tips will help you find useful websites more quickly:

- Know exactly what you want to find out about first.
- Use only a few important keywords in a search, putting the most relevant words first.
- Be precise. Only use names of people, places, or things.

Disclaimer

All the internet addresses (URLs) given in this book were valid at the time of going to press. However, due to the dynamic nature of the Internet, some addresses may have changed, or sites may have ceased to exist since publication. While the author and publishers regret any inconvenience this may cause readers, no responsibility for any such changes can be accepted by either the author or the publishers.

Glossary

adapt   develop special features that enable an organism to survive in a particular habitat

allele   form of a gene

amino acid   building block of proteins

antibiotic   special drug that destroys bacteria inside the body

artificial insemination (AI)   transfer of sperm, which has been collected from a male animal, to the reproductive system of a female

asexual reproduction   production of offspring identical to its single parent

autosome   chromosome that carries information about the body cells

bacteria   type of micro-organism that can be helpful, but that can also cause disease

base   building block of the DNA molecule. There are four bases, called adenine, cytosine, guanine, and thymine.

biochemist   scientist who studies the chemistry of life

blood transfusion   process where blood is taken from one person and put into someone else who needs it

cancer   disease caused by the uncontrolled growth of cells

carcinogen   chemical that can cause cancer

carrier   individual who carries a recessive allele for a genetic disease

chromosome   thread-like structure found within the nucleus of cells

clone   genetically identical copy

conception   process of becoming pregnant

cross-breed   use genetic material from two different varieties, or breeds, of plant or animal to create offspring

cutting   piece of a plant that is removed and planted elsewhere, to grow into a new plant

cystic fibrosis   genetic disease that causes thick, sticky mucus to build up in the lungs, digestive system, and reproductive system

cytoplasm   jelly-like substance which fills the cell and in which the components of the cell are suspended

diabetes   disorder which commonly causes an inability to process sugars due to a lack of the hormone insulin

DNA (deoxyribonucleic acid)   molecule that carries the genetic code. It is found in the nucleus of the cell.

dominant   dominant characteristic occurs even when only one allele is present in the allele pair

donor   person who gives something, for example an organ, to another

embryo   baby at a very early stage of development inside the mother

enzyme   protein molecule that changes the rate of chemical reactions in living things without being affected itself in the process

ethical   relating to principles of right and wrong

fertilize   joining together of a male and female sex cell

foetus   unborn baby more than eight weeks into development

gamete   sex cell

gene therapy   curing genetic diseases by altering the genes

gene   individual unit of information on the chromosome

genetic cross   when two genetically different parents create offspring

genetic diagram   way of representing the transfer of genes from parents to offspring

genetic engineering   technique for changing the genetic information in a cell

genetic information   information contained in the chromosomes, within the nucleus of each cell

geneticist   scientist who studies genetics

genotype   alleles that make up the genes of an individual

germ line therapy   process carried out on the eggs and sperm so that genetic disease is completely removed and cannot be passed on

habitat   place where an animal or plant lives

haemophilia   genetic disease that affects males and prevents the blood from clotting properly

heterozygous   term used when alleles in a pair are different

homozygous   term used when both alleles in a pair are the same

hormones   chemical messages carried around the body in the blood

human genome   sum of all the genes on all the chromosomes of a human being

inherit   pass something on from parents to offspring

insulin   a hormone made in the pancreas involved in the control of blood sugar levels

ionising radiation   particles or rays given off by radioactive nuclei

karyotype   special photograph of human chromosomes

malaria   tropical disease spread by mosquitoes

micro-organism   bacteria, viruses, and other minute organisms which can be seen only using a microscope

miscarriage   when a pregnancy fails to develop and the foetus is lost

molecule   group of atoms bonded together

moral   relating to principles of right and wrong

mutation   change in the genetic material

natural selection   survival of the fittest organisms, and the passing on of their genes through reproduction

nervous system   system including the brain, spinal cord, and nerves, that controls the body's responses to things

nucleus   central part of the cell, the nucleus contains the DNA

organism   individual living thing, such as a plant or animal

ova   female sex cell in animals

ovule   female sex cell in plants

pancreas   organ that produces the hormone insulin

phenotype   what an individual looks like

pollen   male sex cell in plants

protein   important building block of living things

recessive   recessive characteristic only occurs if the allele pair is made up of two recessive alleles

red blood cell   blood cell that carries oxygen around the body

reproduction   producing new individuals of the same type of organism

semen   fluid that contains the sperm

sex chromosome   chromosome that determines the sex of the offspring

sexual reproduction   involves the joining of special male and female sex cells to form an individual that is different from both its parents

sickle cell anaemia   genetic disease that causes the red blood cells to have a strange shape which causes them to carry less oxygen

species   specific group of very closely related organisms whose members can breed successfully to produce fertile offspring

sperm   male sex cell in animals

transgenic   animal or plant that contains a gene from an organism of a different species, put in place by genetic engineering

ultraviolet light   rays that are beyond the visible light spectrum

uterus   part in the female where a developing baby forms

X chromosome   sex chromosome in humans that carries information about being female

xenotransplantation   transplanting an organ from one species of animal into another completely different species

Y chromosome   sex chromosome in humans that carries information about being male

Index

Titles in the *Life Science in Depth* series include:

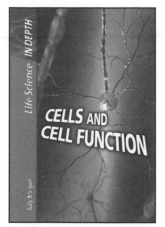

Hardback   0 431 10896 X

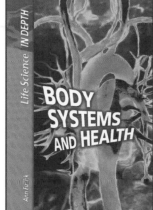

Hardback   0 431 10897 8

Hardback   0 431 10898 6

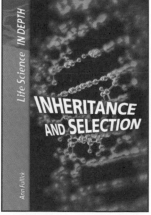

Hardback   0 431 10899 4

Hardback   0 431 10900 1

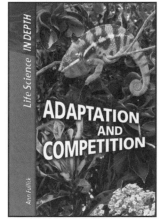

Hardback   0 431 10901 X

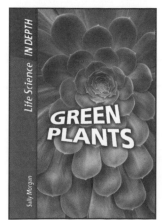

Hardback   0 431 10910 9

Find out about other titles from Heinemann Library on our website www.heinemann.co.uk/library